Food
81

疯狂的蟋蟀
Crazy Crickets

Gunter Pauli

[比] 冈特·鲍利　著

[哥伦] 凯瑟琳娜·巴赫　绘

李欢欢　牛玲娟　译

<figure>上海远东出版社</figure>

丛书编委会

主　任：田成川

副主任：何家振　闫世东　林　玉

委　员：李原原　翟致信　靳增江　史国鹏　梁雅丽
　　　　任泽林　陈　卫　薛　梅　王　岢　郑循如
　　　　彭　勇　王梦雨

特别感谢以下热心人士对童书工作的支持：

匡志强　宋小华　解　东　厉　云　李　婧　庞英元
李　阳　刘　丹　冯家宝　熊彩虹　罗淑怡　旷　婉
杨　荣　刘学振　何圣霖　廖清州　谭燕宁　王　征
李　杰　韦小宏　欧　亮　陈强林　陈　果　寿颖慧
罗　佳　傅　俊　白永喆　戴　虹

目录

Contents

一只蟋蟀上蹦下跳的，看起来很不安。一只黑猩猩很想吃蟋蟀这种高蛋白的食物，并且很想知道蟋蟀被抓住后会想些什么，他害怕成为灵长类动物的食物吗？

"我亲爱的蟋蟀，"黑猩猩调侃道，"你知道吗？你可是我营养美味的大餐噢。"

A cricket is running up and down looking very worried. A chimpanzee, keen on eating this concentrated form of protein, wonders what idea the cricket has got hold of now. Could it be that it is scared to form part of a primate's diet?

"So, my dear cricket," teases the chimp, "are you aware that you are a really rich, nutritious dinner for me?"

你可是营养美味的大餐噢

you are a rich nutritious dinner

大家都应该吃昆虫

Everyone should be eating insects

"我知道，似乎到处都在流传说，大家都应该用吃昆虫代替吃肉。"

"人类现在才知道这一点，真令人惊讶。我们早就知道你富含蛋白质了。"

"是的，但至少两条腿行走的人类现在已经清楚认识到，他们不能继续通过吃动物来填饱肚子了。"

"I know, it seems that the buzz has gone around – that instead of eating meat, everyone should be eating insects."

"It is surprising that people only figured this one out now. We have known for years that you are a protein bomb."

"Yes, but at least those walking on two legs are now well aware that they can't continue to eat animals to fill their empty stomachs."

"蟋蟀是最富含蛋白质的食物来源之一。如果你和你的兄弟姐妹同意被吃掉，可以挽救许多动物的生命。"

"这不公平。难道只有付出我的生命，才能挽救他们的生命吗？"

"嗯，你和你的家族成员繁殖很快，还吃别人不吃的食物，所以你们是食物链的关键部分。如果你们能一直存在，每个人就可以有足够的食物来源。"

"Crickets are one of the richest sources of protein. If you, and your brothers and sisters, will agree to be eaten, it could save the lives of many animals."

"That is not fair. I have to pay with my life so they can have theirs?"

"Well, you and your families can reproduce so fast and eat what no one else will eat, so you are a key part of the food chain. If you play along, everyone could finally have enough to live on."

你繁殖很快

You can reproduce so fast

我愿意成为大家的点心

I will agree to be turned into a snack

"你说得对，我们确实繁殖得非常快，如果能让我的孩子们生活得更幸福，我愿意成为大家的点心。"

"我们很钦佩你能有这种生命循环的想法！"

"You are right, we do reproduce very fast, and I will agree to be turned into a snack if it means my kids would live more happily."

"We appreciate that you have such a circular vision of life!"

"我想知道，为什么人们要从原来四条腿行走变成两条腿直立行走？用四条腿跑得更快啊！"

"是的，四条腿行走会更稳一些，还有助于提高在地面的移动速度。但如果你用两条腿行走，一个优势是你可以腾出两只胳膊携带食物了。"

"I wonder why people, who could once run on four legs, decided to stand up and walk on two. You can run faster on four!"

"Yes, running with four legs is more stable, and gives better ground speed. But there is one advantage: if you run on two legs then you have two arms free to carry food."

如果用两条腿行走你可以携带食物

If you run on two legs you can carry food

可以和家人一起分享食物

you can share food with your family

"携带食物有什么用？你还是要吃掉啊！"

"是的，但是如果你用四条腿行走的话，就不能边走边吃了。"

"有意思！所以你现在可以边行走边携带食物，那又怎么样呢？"

"想想啊，这样你就可以回家和家人一起分享食物了！如果你是一个好猎手，能找到更多的食物，你就可以给家人带回去更多。"

"What is the use of carrying food? You have to eat it!"

"True, but if you need four legs to get around, you can't run and eat at the same time."

"Interesting! So now that you can walk and carry food, what then?"

"Come on, now you can get home and share food with your family! If you are a good hunter and find more food, then you can bring a lot of it with you."

"你的意思是，人类——好
斗的智人——会愿意分享吗？"

"嗯，不仅是合作与分享，而且是想活得
更好。"

"你确定吗？人类是好斗的，会强烈捍
卫他们的领土——只存在于脑海中
的一个假想边界。"

"You mean that man, this
aggressive Homo sapiens, is
prepared to share?"

"Ah, not only cooperate and share, he
is better now prepared to survive."

"Are you sure? Human beings can
be aggressive, and fiercely defend
their territory, based on the illusion
of a border that only exists in their
mind."

人类乐于分享

Homo sapiens is willing to share

分享创造了爱与关心！

Sharing leads to loving and caring!

"好吧，你说得没错，我们都有过一些不好的经历。但我们都知道，当你带着食物回家时，家人会喜欢你。如果你天天都带着食物回家，他们会与你更加亲密。"

"你是说正因为做到了分享，从而得到了爱与关心这样的回馈？"

"Well, you are right and we all have lived through a few bad experiences, but you and I both know that when you come home with food, family do like you. And, if you come home with food every day, they will bond with you."

"You mean that there is in fact a reward, such as loving and caring, because you are sharing?"

"事实的确如此！如果你多与家人以及周围的人分享，就可以形成一个群体，即使在困难的时刻也相互关心。"

"我很快就会被吃掉了，这岂不是很遗憾吗？"

……这仅仅是开始！……

"In fact, there is! And, if you share more with your family and others around you, then you can build a community where people care for each other – even when the going gets tough."

"Isn't it a pity that I am going to be eaten soon?"

... AND IT HAS ONLY JUST BEGUN!...

……这仅仅是开始！……

... AND IT HAS ONLY JUST BEGUN! ...

Did You Know ?

你知道吗?

The cradle of humankind is in Africa. Biologically speaking there are no races. We are all Africans. It is here where people learned about sharing and cooperation millions of years ago.

人类起源于非洲，从生物学角度讲没有种族之分，我们都是非洲人。几百万年前人类在这里学会了分享与合作。

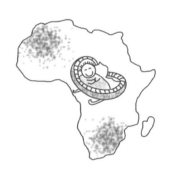

An adult brain is a forest of neurons that contains 100 billion nerve cells that have more than 100 trillion connections. Unfortunately we only use 10% of its potential.

成人的大脑包含1000亿个神经细胞和100万亿个神经突触，它们相互连接。可是，人类只开发了其中10%的潜能。

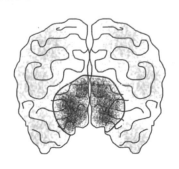

Two billion people worldwide eat insects on a regular basis: in Africa, Asia, Latin America and Australia, but not in Europe or America.

全世界有20亿人经常吃昆虫，他们分布在非洲、亚洲、拉丁美洲和澳大利亚，但是在欧洲和美国没有分布。

蛋白质

昆虫吃掉2千克的饲料能生成1千克的蛋白质，这种蛋白质含有的铁元素比牛肉更丰富。更好的是，昆虫不需要专门的土地，它们吃剩下的食物，减少了温室气体的排放。

Insects eat two kg of feed to produce one kg of protein, and this is richer in iron than beef. Even better, insects require no land; they eat leftovers and emit fewer greenhouse gases.

Roasted worms have a nutty flavour, while mealworms take on the taste of their last meal. So, if they have been eating apples, they will taste like apples.

烤蠕虫有坚果的味道，而面包虫的味道和它最后吃的东西相似。所以，如果面包虫吃过了苹果，它们的味道尝起来就会像苹果。

Insect protein relieves the pressure on making fish meal for aquaculture and to feed livestock. This can reduce over-fishing. One ton of insect protein saves one ton of fish.

使用昆虫蛋白可以缓解水产养殖和家畜养殖所需饲料的压力，也可以防止过度捕捞。1吨昆虫蛋白可以替代1吨鱼类的蛋白。

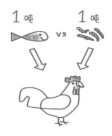

有1900种常见昆虫被人类食用，最受欢迎的种类有甲虫、毛毛虫、黄蜂、蚂蚁、蚱蜢、蝗虫和蟋蟀。

There are 1,900 popular insects eaten by humans and the favourite ones are beetles, caterpillars, wasps, ants, grasshoppers, locusts and crickets.

1千克的蟋蟀粉由18 000只蟋蟀制成，蟋蟀在短短的6周时间内就能发育成熟。

One kg of cricket flour contains 18,000 crickets, who grew to maturity in just 6 weeks.

How do you feel about having insects for breakfast?

你觉得早餐吃昆虫怎么样?

如果有人能劝你吃不健康的肉，其他人还能说服你吃健康的昆虫吗?

If someone convinced you to eat unhealthy meat, could someone else convince you to eat healthy insects?

Are you happy when someone in your family prepares food for everyone when they come home for meals?

回家吃饭时，家人已经为大家准备好了食物，你高兴吗?

你喜欢人类起源于非洲这个说法吗? 为什么?

Do you like the idea that we are all from African origin? Why or why not?

Try to run on all fours, using your legs and arms. It does not work well, does it? We have no rhythm and we lack the right body posture and even feel uncomfortable. Still, give it another try. The experience gives you a sense of the transformation humans went through over a long time. Now, try to carry something while you are on all fours. The only place where you could perhaps keep anything is in your mouth. You will experience the power and the logic of walking and carrying something at the same time. It was not only a smart thing to do,it was necessary.

　　尝试四肢着地用腿和胳膊移动，好像不是很协调，是不是？这样做节奏不对，身体姿势不对，甚至会感到不舒服。不过，再试一次，这会让你感受到人类漫长的变化。试着趴在地上携带一些东西，这时身上唯一能携带东西的部位就是嘴了，你将会感受到边行走边携带东西的力量。这不仅是一种聪明的行为，而且是一种有必要的行为。

学科知识
Academic Knowledge

生物学	可食用昆虫包括甲虫、毛毛虫、蜜蜂、黄蜂、蚂蚁、蚱蜢、蝗虫、蟋蟀、蝉、飞虱、白蚁、蜻蜓和苍蝇；人类大脑进化了600万年后停了下来；人脑由大脑、小脑和脑干三部分组成；动脉为大脑输送体内25%的血液，当你思考问题时大脑会消耗体内50%的氧气。
化 学	大脑会被神经毒素损伤；一些物质如铅（添加到汽油和油漆里的）和乙醇（添加到燃料和饮品里的）会破坏神经组织；科伊桑岩石艺术所用的颜料成分是人的尿液、血液和赭石、木炭、鸵鸟蛋，在一万年后仍会保持亮度。
物 理	记忆信号以微电流形式穿过神经细胞，当电荷到达神经突触（细胞连接处）时触发化学物质（神经递质）的释放。
工程学	在地面依靠四肢行走或直立行走；四条腿行走可以保证更快的速度和更高的稳定性，两条腿行走可以发挥手的作用，并把头抬到地面上以呼吸凉爽的空气。
经济学	价格由供需平衡和人们的期待值决定；昆虫粉的价格和小麦粉、玉米粉相近，但是它的蛋白质含量比更贵的鱼肉还要高。
伦理学	利他主义认为从道德出发，个体有义务帮助他人。
历 史	非洲南部遗传变异显示，现代人类是非洲南部早期人类混合的结果；洞穴壁画表明科伊桑人生活在2万至3万年以前。
地 理	"人类摇篮"遗址位于约翰内斯堡附近，是联合国教科文组织世界遗产之一，这里发现的人类祖先化石占全球的三分之一；南非是拥有石器时代雕塑、绘画和雕刻收藏品最多的国家。
数 学	智商高低与物种的脑体积有关；绝对数与相对数的区别。
生活方式	如果你付出的代价是你无法承受的，那是一种牺牲；父母为自己的孩子们作出牺牲，通常是为了让自己的孩子们实现自己未完成的梦想。
社会学	科伊桑人具有非暴力和愈合文化，他们相信自己生活在"愈合的土地上"，他们对着星星唱歌，几乎不发动战争；循环或线性的时间概念决定着人们是否相信轮回转世；社会神经学显示，人们对与他人保持联系的需求比对食物和住所的需求更迫切。
心理学	人们相互分享是因为他们彼此关心，希望能保持联系；礼物是利他主义的爱、赞赏和感激的普遍表达；理解别人的想法，明白别人的希望、恐惧和动机，能让我们生活得更和谐，却也约束自身的更好发展；为什么一些人可以很快从挫折中恢复而另一些人却不能，这与语境敏感性、社会直觉、应变能力和自我认知相关。
系统论	人类进化的方式与分享经历相关；人类社区随着时间推移而关系加深，这使我们看到完整的生活体系。

情感智慧
Emotional Intelligence

蟋蟀

蟋蟀感受到压力，表现出了很多焦虑，不知道在人类对昆虫蛋白的需求中该如何存活下来，但他仍然保持清醒的头脑，看到不仅是他，而且是整个世界都处于危险当中。蟋蟀反对这样的观点——为解救动物生命以及保存足够多的食物，自己必须死去。蟋蟀为了孩子们的未来，最终接受了自己死亡的归宿，认为这是自己在改造世界中发挥的作用。在经历了这个艰难的选择后，蟋蟀开始提出关于人类进化的基本问题。蟋蟀的问题很有逻辑性，他根本不信任人类，不认为人类会与他人分享任何利益。但在听了黑猩猩的解释后，他看到了建立强大生物群体的可能性。但接着又要直面现实问题：很快就会被吃掉了吗？

黑猩猩

黑猩猩很自信，用黑色幽默的方式调侃蟋蟀，提醒蟋蟀可能会成为他的食物，同时也对蟋蟀富含营养表示感激。黑猩猩正确地理解蟋蟀的牺牲：拯救了其他动物的生命，为自然界提供了更充足的食物来源。黑猩猩与蟋蟀一起互动，解释了人类随着时间是如何进化以及人类直立行走的原因。另外，黑猩猩深入探讨群体的形成，何种分享是可以得到回报的。当蟋蟀存有疑虑时，黑猩猩提供了清晰的思路，让蟋蟀知道何时需要作出牺牲。

艺术
The Arts

看一看科伊桑岩画，它们至少有1万年的历史了。岩画本身不是关键，颜料为什么能保持这么长时间才是关键。你想画这样的画吗？查找颜料成分，尝试自己去画。如果科伊桑人上万年前都可以做到，你同样也可以做到。不要惊讶颜料的成分，它们可能是一些你意想不到的东西！

思维拓展
Systems: Making the Connections

人类出现在地球上之后，数量不断增加，超出了自然界的承载能力。而且，随着人类饮食结构的转变，问题日益凸显。人类更想吃动物蛋白的原因是复杂的，实际上，以谷类、豆类和鱼类为食物来源的鸡、猪和牛，可以直接提供给人类营养物质。利用食物生产更多的食物，是食物链效率低，以至于不能满足人类对营养物质基本需求的主要原因。人类人口增加了几十亿，这使得形势越来越严峻。虽然科研人员在试图寻找食物增产的方法，但现在应该是时候改变策略了：在可持续生产上做一些事情。如果人类消耗的食物不仅仅可再生，也更好、更健康，这样我们就可以有更好的生活。许多反对者可能会不喜欢"其他食物"，或者是"习惯"很难改变。然而，在过去几十年中我们周围的食物已经改变。如果我们有那么多人愿意吃转基因玉米，或者接受吃含激素的汉堡包，那么能不能想象一下，把昆虫作为地球上70亿人口的食物呢？

动手能力
Capacity to Implement

列出全球最受欢迎的10种昆虫，问问自己哪种是可以吃的。然后，查看所有可以添加昆虫粉的食谱，选出自己喜欢的食物，看看当地有没有可以用的食材。如果有，尝试做一下。如果没有，你准备好自己动手去养殖这些昆虫了吗？

故事灵感来自

This Fable Is Inspired by

玛丽·瑞德
Marie Ryd

　　玛丽·瑞德在瑞典卡罗林斯卡学院学习并获得医学博士学位，专注于心理学和企业领导力心理学方法的研究。在成为科学记者之前，她在牙科实习过，在细菌学和免疫学学科上做过研究，这促成了 Holone 杂志的创立。Holone 杂志致力于脑研究的最新发现，尤其专注于社会行为对大脑影响的研究。

图书在版编目（CIP）数据

冈特生态童书.第三辑修订版:全36册:汉英对照 /
（比）冈特·鲍利著;（哥伦）凯瑟琳娜·巴赫绘;
何家振等译.—上海:上海远东出版社,2022
书名原文:Gunter's Fables
ISBN 978-7-5476-1850-9

Ⅰ.①冈… Ⅱ.①冈… ②凯… ③何… Ⅲ.①生态环
境–环境保护–儿童读物—汉、英 Ⅳ.①X171.1-49

中国版本图书馆CIP数据核字（2022）第163904号
著作权合同登记号图字09-2022-0637号

策　　划　张　蓉
责任编辑　程云琦
封面设计　魏　来　李　廉

冈特生态童书

疯狂的蟋蟀

[比]冈特·鲍利　著
[哥伦]凯瑟琳娜·巴赫　绘

李欢欢　牛玲娟　译

记得要和身边的小朋友分享环保知识哦！
八喜冰淇淋祝你成为环保小使者！